Return of the TERNS

How Scientists Are Saving Island Birds

by Jennifer Keats Curtis with Kim Abplanalp
illustrated by Phyllis Saroff

This book is about a program that takes place in Maryland. The author wishes to thank the Maryland Coastal Bays Program's Dr. Roman Jesien; Maryland Department of Natural Resources' Dave Brinker; and U.S. Fish and Wildlife's Pete McGowan, without whom, this book could not have been written.

A giant sandbox floats on the water. The massive square is the size of an apartment. It's full of crushed shells, green spiky grass, tiny tents, and birds, so many birds.

The birds are about the size of a hot dog in a bun. With a little black cap on their white heads, they hop about on skinny, orange legs. Their beaks, the color of catsup, open. Shrilly, they call: *keek, keek, keek.*

This sandbox is not easy to find. It's far from nosy neighbors—people and animals. This secret place is Bird Town, or in this case, Tern Town.

These beautiful birds are Common Terns. They are not really in a sandbox. Tern Town is a big, floating island. Scientists created this island (a big raft) as a seasonal getaway. The birds' usual summer home—a small island—has washed away. And, if not for this raft, these terns would have nowhere to go.

The little tents are shelters made of wood. The grass is fake. So are some of the birds.

The still ones are decoys. The scientists use dummy birds to attract the terns to this new spot. Several "*keek keek keeks*" are recordings that the experts use for the same reason. The terns do not seem to care. They treat the floating raft like their normal nesting grounds—sand, crushed shells, and grasses on little islands.

Why do birds fly to the same place every year? The winged creatures have maps (and maybe calendars) in their heads. That chart in their brains steers them to the same migration spots about the same time every year.

As small islands wash away, the birds continue flying to their regular summer places. They don't know these sites no longer exist. If their destinations are gone, they must head elsewhere.

The rafts were built out of desperation, when only 30 pairs of terns returned one year. This year, more than 300 pairs have made the raft their summer home. Until these little islands are rebuilt, scientists hope the birds will use artificial islands as habitats. Otherwise, these animals have nowhere to stop, rest, lay eggs, and bring up their families. They will remain locally endangered... or worse.

Since it's June, grey, speckled eggs, no bigger than marshmallows, are among the shells. Fortunately, the raft provides more protection than their "normal" island homes.

Most terns scrape out a little hole with their feet and beak. They wiggle their body into the sand and shells to make a shallow nest. This means their eggs are more likely to get wet during high tides. They are easily found—and eaten—by predators, like gulls and birds of prey.

On beaches, the eggs are well camouflaged. Well-meaning people and their dogs easily crush them, often without knowing it.

Now, seeing so many birds using this raft is exciting; but it's not enough. The scientists will know the raft is working if the terns return each spring. Can the rafts help save these birds until a more permanent solution is found?

Each week, the scientists boat to the raft to band the terns. The birds get two pieces of jewelry—a metal anklet on one leg and a white band with black numbers and a letter on the other. These show where the birds are from and let scientists read the lettering from afar.

To catch the adult birds, scientists first pick up the eggs and place them in a safe container. Then they place fake eggs in the nest and cover it with a small, clear trap.

When a tern plops down on the fake egg to incubate it, the trap closes. The birds don't seem to notice. That is, until they are carefully scooped up in a gentle hand.

The scientists measure the birds' beaks and wing spans. They place the bands around the terns' legs. With their new jewelry in place, they are returned to the raft and the real eggs. Now the scientists can easily identify raft returnees next year.

Within weeks, the babies hatch, using a little egg tooth to break out of their eggs. The chicks are covered in down. Their eyes are open. They can even walk.

When the fluffy chicks are a few days old, the scientists put on bike helmets and hard hats and return to band them. This headgear protects the scientists from unhappy tern parents that divebomb anyone who gets near their babies!

Most hatchlings won't return to the raft until they are old enough to breed and have families of their own.

By fall, the birds have flown the coop, (well, the raft), and are on their way back to South America.

Once all of the birds are gone, the scientists tow the raft back to land. They'll roll it and haul it up the ramp to store it for winter. In the spring, they will lug it back to the secret spot and redecorate it with shells, grasses, small shelters, and decoys.

Then, they'll await the return of the terns.

For Creative Minds

Match the Terns by Age

Identify the tern by its age.

1.   I am an egg waiting to hatch.

2.   I am a hatchling, just breaking out of my egg.

3.   I am waiting for my brother or sister to hatch. We are well cared for by both parents.

4.   I am four or five days old, a young nestling.

5.   I am a fledgling and can fly.

6.   I am an adult.

A

B

C
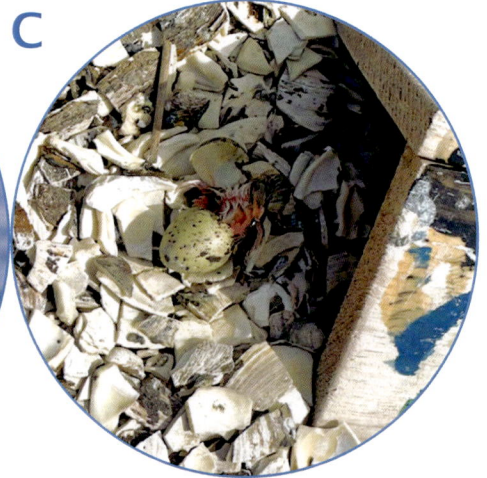

D

E

F

Answers: 1D; 2C; 3F; 4A; 5E, 6B

Shorebird Snacks

Different shorebirds eat different foods. Common Terns eat small fish including American sand lance, menhaden, spot, Atlantic silverside, bay anchovies, and occasionally sand crabs! Find what the terns are eating in the photographs.

Fun Facts

Match the statement to the photo.

1. Terns cool themselves down by "panting." Since birds can't sweat, they cool off by opening their beaks. They may also spread their wings to feel the breeze or take a dip in the water.

2. Adults make a nest, called a scrape, by moving the sand (or broken shells) with their beaks and then wiggling their bellies until they are settled in. Sometimes, they add sea grasses that blow onto the raft.

3. Both adults and chick terns are banded to allow scientists to recognize individuals, record information about them, and report when the terns leave the raft.

4. The raft is made of sections that are assembled, locked together and anchored for the terns' summer nesting season. When put together, the raft is 48 x 48 feet.

5. The Maryland's Coastal Bays Program hosts a "Bay Day" each year where children can paint the shelters to be used on the raft.

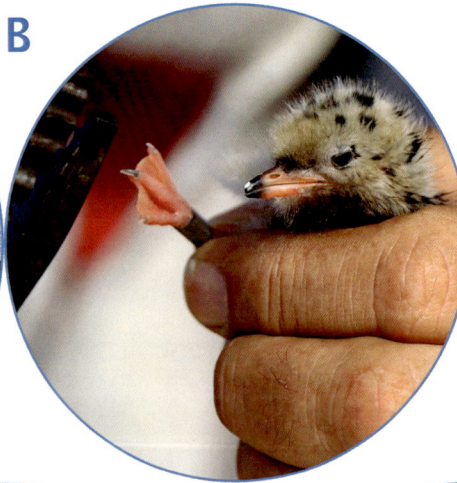

A

B

C

D

E

Answers: 1E; 2C; 3B; 4A, 5D

What's on the Raft?

Everything the scientists put on the raft has a reason. See if you can match the description to the item.

1. Shelters give the birds a place to get out of the direct sun and be safe from predators.

2. Decoys help the birds think that there are already other birds on the raft, letting them know it's a safe place. Scientists also play a lure to attract the birds.

3. Artificial grasses are "planted" because the scientists would not be able to water real grasses on the raft in the bay.

4. Broken shells are used instead of sand that may get blown off by wind.

5. Young birds might be strong enough to fly off the raft but not yet grown enough (or muscular enough) to lift off of the water. The birds can paddle over to the ramp to walk up to the raft. The ramp's slope is the same as the slope of the island.

Answers: 1D; 2C; 3A; 4E, 5B

The raft in this book is based on the work of a partnership between Maryland Coastal Bays Program (MCBP,) Maryland's Department of Natural Resources (DNR) Wildlife Heritage Service, and Audubon Mid-Atlantic. They wish to thank their terrific group of volunteers who make this project a success. Special thanks to Todd Peterson, John Collins, Karin and Tom Johnson, and Frances and Matt Cole for all their support.

Library of Congress Cataloging-in-Publication Data

Names: Curtis, Jennifer Keats, author. | Abplanalp, Kim, author. | Saroff,
   Phyllis V., illustrator.
Title: Return of the terns : how scientists are saving island birds / by
   Jennifer Keats Curtis with Kim Abplanalp ; illustrated by Phyllis
   Saroff.
Description: Mt. Pleasant, SC : Arbordale Publishing, LLC, 2025. | Includes
   bibliographical references.
Identifiers: LCCN 2024033507 (print) | LCCN 2024033508 (ebook) | ISBN
   9781638173298 (trade paperback) | ISBN 9781638173335 (ebook) | ISBN
   9781638173373 (epub) | ISBN 9781638173410 (pdf)
Subjects: LCSH: Terns--Habitat--Juvenile literature. | Bird
   refuges--Maryland--Juvenile literature. | Wildlife
   conservation--Maryland. | CYAC: Terns. | Habitat (Ecology) | Bird
   refuges. | Wildlife conservation. | Maryland.
Classification: LCC QL696.C46 C87 2025 (print) | LCC QL696.C46 (ebook) |
   DDC 333.9581609752--dc23/eng/20240824
LC record available at https://lccn.loc.gov/2024033507
LC ebook record available at https://lccn.loc.gov/2024033508

English Lexile® Level: 830L

Bibliography
A "Hail Mary" to Save Some Species of Birds in Maryland." PBS NewsHour, 6 June 2021, www.pbs.org/newshour/show/a-hail-mary-to-save-some-species-of-birds-in-maryland.
"A Tale of Two Colonies." Conserve Wildlife Foundation of New Jersey, 24 Sept. 2015, www.conservewildlifenj.org/blog/2015/09/24/a-tale-of-two-colonies/.
Admin, Page. "Endangered Terns Nest on Raft Built by Md. Coastal Bays - Worcester County News Bayside Gazette." Berlin, Ocean Pines News Worcester County Bayside Gazette, 13 Apr. 2023, baysideoc.net/endangered-terns-nest-on-raft-built-by-md-coastal-bays/.
Bon Voyage to the Terns! - OceanCity.com. 13 Sept. 2022, www.oceancity.com/bon-voyage-to-the-terns/.
"Common Tern Recovery Project." Audubon Vermont, 21 Jan. 2016, vt.audubon.org/conservation/common-tern-recovery-project.
"Common Tern Sounds, All about Birds, Cornell Lab of Ornithology." Www.allaboutbirds.org, www.allaboutbirds.org/guide/Common_Tern/sounds#:~:text=In%20flight%20or%20in%20territorial.
"Could Artificial Islands Be the Key to Saving Some Endangered Birds?" Scripps News, scrippsnews.com/stories/could-artificial-islands-be-the-key-to-saving-some-endangered-birds/.
Liptak, Matt. "2022 Maryland Common Tern Colony Saw Explosive Growth Thanks to Raft Project." Maryland Wilds, 16 Oct. 2022, marylandwilds.com/2022/10/16/2022-maryland-common-tern-colony-saw-explosive-growth-thanks-to-raft-project/.
Magazine, Hakai. "Threatened Seabirds Get a Life Raft in Maryland." Hakai Magazine, hakaimagazine.com/news/threatened-seabirds-get-a-life-raft-in-maryland/.
"Nesting Platform Initiative for Endangered Birds in Maryland Coastal Bays Is a Big Success." News.maryland.gov, news.maryland.gov/dnr/2022/09/29/nesting-platform-initiative-for-endangered-birds-in-maryland-coastal-bays-is-a-big-success/.
Ralph Simon Palmer. A Behavior Study of the Common Tern. 1941.
"Restoring the Shore: Touring Coastal Projects in Maryland's Coastal Bays Region | Maryland Sea Grant." Www.mdsg.umd.edu, www.mdsg.umd.edu/onthebay-blog/restoring-shore-touring-coastal-projects-marylands-coastal-bays-region.
Wendy Coolen. "Beachnester Buzz: A Day in the Life of a Beachnester." Conserve Wildlife Foundation of New Jersey, 8 Aug. 2016, www.conservewildlifenj.org/blog/2016/08/08/beachnester-buzz-a-day-in-the-life-of-a-beachnester/.
"Will a Few Good Terns Attract Others?" WYPR, 30 July 2021, www.wypr.org/wypr-news/2021-07-30/will-a-few-good-terns-attract-others.

Printed in the US
This product conforms to CPSIA 2008

Arbordale Publishing, LLC
Mt. Pleasant, SC 29464
www.ArbordalePublishing.com

Snag It!
Who Needs a Dead Tree?

by Mary Holland

Snag It!

Who Needs a Dead Tree?

Explore the crucial role of dead or dying trees, known as snags, in forest ecosystems. Through engaging narrative and vivid photography, the book explains how snags provide essential habitats for a variety of animals, including birds, mammals, reptiles, and insects. It highlights the diverse ways animals use snags for nesting, shelter, food, and perching. Ideal for environmental education and aligned with NGSS standards, this book promotes wildlife conservation and ecological awareness to young readers.

Arbordale Publishing offers so much more than a picture book. We open the door for children to explore the facts behind a story they love.

Animals featured in this book using snags include bald eagle, bat, flying squirrel, gartersnake, great blue heron (cover), green heron, porcupine, raccoon, red squirrel (title page), red-breasted nuthatch, salamander, screech owl, turkey vulture, and woodpecker.

The *For Creative Minds* includes
· Pileated Woodpecker Food and Nest Holes
· Which Animals Use Snags?
· How Might These Animals Use Snags?
· Nests in Snags
· True or False?

Arbordale's interactive ebooks read aloud in both English and Spanish with word-highlighting and adjustable audio speed. Available for purchase online.